The Beauty of Australian Wildflowers

The Beauty of Australian Wildflowers

JOHN BROWNLIE
SUE FORRESTER

VIKING O'NEIL

Viking O'Neil
Penguin Books Australia Ltd
487 Maroondah Highway, PO Box 257
Ringwood, Victoria 3134, Australia
Penguin Books Ltd
Harmondsworth, Middlesex, England
Penguin Books
40 West 23rd Street, New York, N.Y. 10010, U.S.A.
Penguin Books Canada Ltd
2801 John Street, Markham, Ontario, Canada L3R 1B4
Penguin Books (N.Z.) Ltd
182-190 Wairau Road, Auckland 10, New Zealand

First published by John Currey, O'Neil Publishers Pty Ltd 1982
as *Australian Wildflowers*
This edition published by Penguin Books Australia 1987
Copyright © Photographs: John Brownlie, 1982
Copyright © Text: Susan Forrester, 1982

All rights reserved. Without limiting the rights under
copyright reserved above, no part of this publication may
be reproduced, stored in or introduced into a retrieval system,
or transmitted, in any form or by any means (electronic,
mechanical, photocopying, recording or otherwise) without
the prior written permission of both the copyright owner
and the above publisher of this book.

Produced by Viking O'Neil
56 Claremont Street, South Yarra 3141, Australia
A Division of Penguin Books Australia Ltd

Designed and typeset in Australia
Printed and bound in Singapore through Bookbuilders Ltd

National Library of Australia
Cataloguing-in-Publication data

Brownlie, John.
  The beauty of Australian wildflowers.

  Includes index.
  ISBN 0 670 90049 4.

  1. Wildflowers – Australia – Identification.
  I. Forrester, Susan Glen. II. Title.
  III. Title : Australian wildflowers.

582.13'0994

Half title page: **Waratah** *Telopea speciosissima* A member of that diverse family, Proteaceae, the Waratah is a masterpiece of intricate structure and glorious colour. Confined to the sandstone regions of New South Wales, it is not common and, unless in bloom, may be overlooked by the casual observer. However, once attuned to its form, it is surprising how quickly one may recognise it. Generally, there will be only one or two at a time; sandstone ledges and shelves are favoured sites, where a little moisture may collect.

Opposite title page: Suspended in time and space, a still blue tarn lies protected within Tasmania's Cradle Mountain Lake St Clair National Park. Such tarns are relics of long-gone glacial periods. Often fairly shallow, their edges are matted with a profusion of interwoven small herbs, heaths and grasses.

Title page: **Billy Buttons** *Craspedia* species Affectionately known as Billy Buttons, a number of *Craspedia* species occur in the high country. Several have not yet received botanical names. Typically, they grow as clumps, with soft woolly leaves and stems of silver or grey-green, topped by heads of gold, cream, yellow or vibrant orange. The slopes and broad valleys of the Kosciusko alpine area are carpeted with the greatest diversity and range of species, and the peak of their flowering occurs from late January to mid-February.

PICTORIAL ACKNOWLEDGEMENT
The author, photographer and the publishers wish to thank Bill Molyneux for the use of ten photographs in this book.

Contents

Introduction	1
Acacias	4
Banksias	8
Eucalypts	13
Grevilleas	18
The High Country	23
The Shoreline	32
Wildflower Forests	38
Sandstone Gardens	49
Arid Australia	63
Glossary	71
Index	73

Introduction

The flora of Australia is vastly different from that of any other continent. About eighty per cent of its species and over thirty per cent of its genera occur nowhere else in the world. No other continent has such a high percentage of unique indigenous flora.

Why is our remarkable flora so different? The answer to this lies in the fact that much of its evolution took place in total isolation from the other continents.

It is widely accepted that a great southern land mass, known as Gondwanaland, existed up to the end of the Jurassic period, about 135 million years ago. Australia was part of this supercontinent, together with South America, Africa, India, Antarctica and New Zealand. Separation of some of these continents began to take place about 120 million years ago, although Australia remained linked to Antarctica until about fifty million years ago.

At the time of Australia's separation from Antarctica, the climate was both warm and moist. Subtropical rainforest vegetation was reasonably uniform and continuous, typified by beech (*Nothofagus*), and conifers, together with members of the Myrtaceae and Proteaceae families, both of which are dominant in our flora today.

The best-known genus of Australia's present-day Myrtaceae family is *Eucalyptus*, but it also includes *Kunzea*, *Melaleuca*, *Baeckea*, *Thryptomene* and *Callistemon*, as well as many rainforest genera.

Proteaceae also contains a number of rainforest species, but is perhaps best-known for its banksias, grevilleas, dryandras and hakeas, together with lesser-known petrophiles, isopogons, adenanthos and lomatias.

Australia gradually drifted further north, becoming more and more geographically isolated; climatic changes took place, progressively cooling and drying out and soils slowly weathered and declined in fertility.

In response to these changes, which took place over millions of years, the flora adapted, and the characteristic which today is its most striking and unique feature evolved: *scleromorphy*, or hardening. Many of our major plant groups are 'sclerophyllous', or 'hard-leaved'. Two probable reasons for this characteristic are that it is a means of resisting and surviving dry conditions; and that it is a response to nutrient deficiencies in the soil.

At least one third of the continent is classified as arid, added to which are the northern areas of monsoonal influence, which experience seasonal aridity. As well as this, most of the remainder of the continent receives a relatively low rainfall, and consequently, by far the greater proportion of the flora has had to adapt to a 'dry' environment.

Thus, subtropical and temperate rainforests have survived only in limited pockets of the continent, mainly in the highlands

Deciduous Beech   *Nothofagus gunnii*
An ancient tranquility pervades the Tasmanian landscape, in the areas where man has not yet made his destructive mark. The rare and endemic Deciduous Beech receives a measure of protection in such areas as the Cradle Mountain Lake St Clair National Park. Its full beauty is seen by relatively few people — the golden autumn tones colour the mountain-sides at a time of year when the weather is particularly uncertain. Sudden snowstorms may change the scene from balmy autumn to bitter cold, with mist and rain blotting all from view.

Deciduous Beech   *Nothofagus gunnii*
Encrusted with a beard of grey-green lichen, *Nothofagus gunnii* (opposite) remains as a relic of the past. Ancient members of the Fagaceae family were once widespread across the continent, though few species have survived. Only in isolated pockets of rainforest regions of eastern Australia will they be found, from the Lamington Plateau in southern Queensland, to the cool moist New England Plateau of New South Wales, and highland forests of Victoria, down to Tasmania's mountainous reaches.

It is in these remnants of cool wet rainforests that we may glimpse something of the scene which existed sixty to 200 million years ago. Rainforest developed to a high level of evolution, containing many species long since replaced by plants which adapted as the climate dried out. Only where the climate has remained favourable do we see the perpetuation of such forests, in which the cycle is one of slow but luxuriant growth, gradual decomposition among deep mosses, lichens and fungi, followed by the emergence of new seedling growth.

and along the coast of the eastern freeboard, where soils are deep and fertile, and rainfall and humidity are high.

Tasmania, alone, still retains something of the environment which prevailed in Gondwanaland times. There, rainforests are deep and rich; *Nothofagus* species and ancient conifers such as *Araucaria*, *Dacrydium* and *Podocarpus* species are all to be found, growing in a singularly beautiful and primeval environment. Tasmania's alpine flora includes many genera and species which are endemic to the island, although much similarity exists between its lowland flora and that of the south-east mainland. This is due to periodic land connections which occurred across what is now Bass Strait.

From time to time, over the past fifty million years, Australia's isolation has been broken. During periods of glaciation (which caused the fall of the oceans of the world) land bridges were formed between the Australian continent and the islands of Indo-Malaysia to the north. Some migration of both flora and fauna took place during such periods, thus accounting for the occurrence of some species of eucalypts and grevilleas (both regarded as being 'Australian' in origin) outside the continent. Elements of tropical flora entered Australia by the same means.

Of necessity, this is a simplification of the complex processes which have modified our flora. Many factors, quite apart from the vast time-span involved, have played a part in its evolution. Inundation of the land by both sea and lake systems; volcanic activity; climatic changes which have caused periods of extreme aridity, high rainfall, and glaciation; and physical alteration to the landscape by uplift, tilting and weathering, have all contributed their influences.

Hakea multilineata

Closely related to the genus *Grevillea*, and also a member of the Proteaceae family, the genus *Hakea* contains many species which are well adapted to hot, dry environments. *Hakea multilineata* (opposite) occurs in the southern sandplains region of Western Australia, around Lake Grace. Typically a tall shrub to 3 or 4 metres, it grows in gravelly heaths in association with low banksias, dryandras and grevilleas, together with numerous small heathland and herbaceous plants. Slender linear leaves surround the bright spikes of flower, which can elongate to 10 or 15 cm, and which irresistibly draw the honeyeaters.

Acacias

Wattle time is as much a part of spring in Australia as blossom time in the northern hemisphere. The blaze of golds and yellows in late winter heralds the new life of spring, and gentle, warm spring days are heady with the fragrance of wattle drifting on the breeze.

The name 'wattle' was originally applied by Australia's early settlers, not to an acacia but to an entirely different species, *Callicoma serratifolia*. This is a shrub of sandstone country in New South Wales. Slender-stemmed, with deeply-veined furry leaves, it bears masses of pale yellow fluffy balls. Its whippy stems and limbs were used in the construction of early wattle-and-daub dwellings, and the vernacular, 'wattle', was gradually transferred to the whole genus *Acacia* — no doubt because of the similarity of the flowers.

Acacias are as widespread across the continent as eucalypts, but, perhaps surprisingly, the genus contains nearly twice as many species. Many of these occur outside Australia. It has been estimated that there are more than 1200 species of acacia, with 700 or so native to Australia. They belong to the family Mimosaceae, and are legumes: that is, they bear their seed in pods.

The majority are found in the cooler or drier regions of the continent, with only a few occurring in rainforest, including a woody vine (*Acacia albizioides*) from Cape York Peninsula. Many species are associated with dry sclerophyll forest; and heathlands, particularly in southern Western Australia, are rich in small-statured acacias. Pure stands of single species such as Mulga (*Acacia aneura*) and Brigalow (*A. harpophylla*) characteristically grow inland. The drooping silver foliage and glowing gold flowers of *A. harpophylla* are a spectacular sight in southern Queensland and northern New South Wales. 'Brigalow country', however, does not necessarily denote floral beauty alone, as Brigalow grows on some of the best agricultural land in that region, and has therefore suffered extensive clearing.

Immense variation in habit, foliage and flower occurs in this genus. Prostrate and ground-covering species such as *A. aculeatissima* and *A. brownii* light up the forest floor of dry woodlands with their pale flowers. The mallee and heathlands of eastern and western Australia also support many a mound of gleaming yellow or deep gold; and the cool, moist forests shelter the soft foliage and filmy flowers of taller shrubby species and trees.

Acacias rarely exhibit true leaves, except in their seedling state. These are generally replaced by phyllodes fairly early in the plant's development. A phyllode is simply a modified stem or petiole, either flattened and broadened, cylindrical or needle-like, which is thought to be associated with adaption of the plant to a dry climate. Most species of acacia produce glands on the surface or edge of the leaves or phyllodes. Speculation exists over the purpose of these glands, but they are

Acacia retinodes
Justifiably deserving its reputation of 'ever-flowering', *Acacia retinodes* is fairly widespread in western and central Victoria, with a coastal population in southern Victoria. It is also known from Tasmania and South Australia. Generally it is seen as a small open tree of 5 to 6 metres, growing as a middle-storey species in open woodland. The phyllodes are long, fine and gently drooping, giving the tree a light, graceful appearance. Golden balls, often clustered toward the ends of the branches, are at their perfumed best during spring and summer, but it is rare to see the species without flower at any time of the year.

Sunshine Wattle *Acacia terminalis*
Happily named, this graceful small shrub (opposite) lights up the shadowy forests of Gippsland in Victoria, coastal New South Wales, and Tasmania. Winter-flowering, its pale, fluffy balls are crowded thickly over the top of the shrub, often giving the effect of a layer of sunlight floating one or two metres above the forest floor. As with many acacias, it has found its way into popularity as a garden shrub. The foliage is particularly pretty, for the ferny leaflets vary in colour from crimson, bronze and tan of the new growth, to deep soft green of the mature leaves.

certainly a useful adjunct to identification of each species. Some well-known species which do retain true leaves include *Acacia terminalis*, *A. elata*, *A. drummondii* and *A. decurrens*. All bear their foliage in pinnate or feathery groups.

As with many species of Australian plants, each 'flower' is in fact composed of many minute flowers bundled together in a ball or spike. The fluffiness of each flower head is imparted by the numerous stamens. Imagine the millions of individual flowers borne on a well-grown specimen of *A. baileyana*, the familiar Cootamundra.

There is a tendency to undervalue acacias, blaming them for hayfever, or labelling them short-lived. But how much poorer both the natural and domestic landscape would be if there were no wattles. And how fortunate we are to have such a wide-ranging, adaptable, and beautiful group of such glorious colour and fragrance.

Ovens wattle *Acacia pravissima*

Found in scattered pockets of central and north eastern Victoria, particularly along the Ovens River, and also in ACT and New South Wales, *Acacia pravissima* (above left) has achieved wide popularity as a garden shrub. Dense and compact, it will grow to 3 or 4 metres in height. The crowded phyllodes are smooth and triangular in shape, and flowering is during spring, when the bush may be covered with masses of soft golden balls.

Acacia cultriformis

Found growing naturally on the Western Tablelands of New South Wales, this species displays an open, sometimes arching habit. Smooth silver-grey phyllodes are an unusual triangular shape, and cluster thickly along the stems. Balls of brilliant gold are displayed over several weeks of spring, and these, together with its attractive foliage and habit, have made the species a popular garden plant.

Acacia drummondii
A well-known plant in cultivation, this Western Australian species (above) is widespread in the south-west, growing in forested country, particularly on streamsides. Its leaves are arranged in pairs of smooth leaflets, giving it a soft ferny appearance. The long golden rods, which bloom from late winter to late spring, are borne on long pedicels and are so numerous that the metre-high shrub is almost a ball of yellow. Its specific name honours one of Australia's earliest and most prolific collector-botanists, James Drummond, who explored widely throughout the south western regions in the early 1800s.

Acacia denticulosa
Almost bizarre in its habit and foliage, this unusual member of the genus is one of many eye-catching Western Australian species. Open-branched and growing to 2 metres, its phyllodes are thick, undulating and strongly ridged, with a scattering of small rasping hairs on both surfaces. Its real beauty lies in its dense, curved, brilliant yellow spikes, which are of outstanding intensity, and 6 to 7 cm long.

Banksias

As spectacular flowers, the banksias command perhaps the most attention of Australia's flora. Often confused with proteas (a South African genus), they belong to the family Proteaceae, and are thus linked with grevilleas, hakeas and dryandras, together with many lesser known genera such as *Petrophile*, *Conospermum*, *Lomatia* and *Adenanthos*.

Banksias were named after Sir Joseph Banks, the famous English botanist who accompanied Captain Cook on his journey of discovery in 1776.

There are nearly sixty named species and a number of varieties recorded in Australia, and with the exception of *Banksia dentata* — a tropical species extending to northern islands and New Guinea — they occur nowhere else.

The majority are found in Western Australia, ranging from ground-hugging and creeping species to those of tree size. The variation in habit alone is of interest, but the loveliness of colour and shape of the blooms, the rich nectar and length of flowering make a real contribution to the floral artistry of the bush.

Not surprisingly, the association between banksias and honey-eating birds is strong. Many species rely on pollination by birds, which have, themselves, evolved strong links with our flora. Indeed, one of the real beauties of the natural history of the whole continent is this vital bond of inter-dependence between flora and fauna. Not only birds, but many insects and, in some cases, mammals such as the Dibbler and the Honey Possum (both denizens of southern Western Australia) and the tiny Pigmy Possums of both sides of the continent, sup on nectar from the banksia blooms. As they feed, pollen is transferred from one flower to the next, and so the flowers are fertilised.

Attractive foliage is a special feature of the banksias. Tough and fairly rigid in most species, and often distinctively lobed, the new growth of many of the shrubby or prostrate species is richly coloured, velvety and as beautiful as any flower. The stems of the ground-creeping species such as *Banksia repens*, *B. prostrata*, *B. blechnifolia* and *B. petiolaris* are also densely furred, almost animal-like, and the very tip of each point of new growth is a tiny russet 'paw'.

Globes, spikes, acorn-shapes, pendant heads, cones, cylinders and balls; gold, rusty-brown, blue-green, orange, cream, lemon and grey, scarlet and crimson and purple — what a superb array nature has presented in these flowers.

But their intricate beauty does not end with the flowers. The fruiting cones are still to come. With the fading of the flowering parts (the style, perianth and bracts, hundreds of which make up each banksia flower-head) the seed capsules or follicles develop in the central core or rachis. It was from the fruiting cone of *Banksia serrata* that the 'bad banksia men' of *Snugglepot and Cuddlepie* fame were drawn. Each evil eye and pair of lips was a follicle, imbued with a wickedness never intended by nature.

The follicle is composed of two valves which contain one or two flat, papery, winged seeds. In many species, they remain tightly closed until a bushfire burns the plant. Then the valves part widely and the seed flutters to the ground, germinating rapidly in the ash and gravelly debris.

Perhaps the most attractive of the cones is that of *Banksia laricina*, commonly known as the Rose Banksia — a reference not to the flower but to the arrangement of the follicles. Each projects like a petal to form a globular 'flower' which is more decorative than the true bloom.

Banksias are desirable garden plants, both for their appearance and their bird attraction, and any garden can be adorned with one or several of this genus.

Banksia baxteri
Common around the sandheaths of the Albany-Stirling Range region of Western Australia, *Banksia baxteri* grows as a sturdy shrub to 3 metres. Large, rigid, broadly serrated leaves surround the globular flowers of greenish-yellow and white. As with most banksias, the perianths open gradually, so that the creamy domed top is obvious only on partially opened flowers. The numerous flowers on each head are followed by the tough, woody seed capsules. The specific name honours a nineteenth century botanical collector, William Baxter.

Banksia hookerana
One of Western Australia's loveliest banksias, this species (above) grows as a small tree or large shrub of 2 to 3 metres. Apricot and cream 'acorn' flowers perch at the ends of the densely-foliaged stems from winter through to summer. The unopened flower is a creamy, woolly white; as the perianth tubes open, the smooth golden styles are released, so that the fully opened flower changes entirely to glowing orange. *Banksia hookerana*, named after an early botanist, Sir Joseph Hooker, occurs naturally in open thickets on deep, sandy soil, extending from Eneabba to Mingenew, some 300 km north of Perth.

Banksia blechnifolia
Similar in flower structure to *Banksia prostrata*, though of a deep bronze-pink, *Banksia blechnifolia* bears distinctively lobed leaves. The new growth is particularly lovely: velvet-soft honey-brown fronds curl gently about the flower, their fern-like quality the reason for the specific name.

Banksia spinulosa
Frequently found as a coastal or woodland shrub, *Banksia spinulosa* (left, above) is one of the best-known eastern Australian species. It occurs from Cairns, in Queensland, to the outskirts of Melbourne. The long cylindrical spikes vary in colour from pure gold to honey-gold with black styles, or gold with red styles. The shrubs flower profusely from autumn to early spring, providing a winter feast for honeyeaters. Dark finely-toothed leaves cluster densely all over the shrub, and it is these which have given the specific name, meaning 'spiny leaved'.

Banksia serrata
Named for its serrated or saw-toothed foliage, this species is common along Australia's east coast, extending from southern Queensland to Victoria, with one isolated occurrence in northern Tasmania. Large silvery-grey and yellow flowers will be seen from spring to early autumn, and are much loved for their nectar by Friarbirds, Little Wattlebirds and Yellow-winged Honeyeaters. The spent flowers are followed by the development of woody follicles, or seed capsules, part buried among the dried styles and each containing one or two black papery seeds. Depending on its situation, it may grow as a dense shrub, even prostrate in very exposed sites, or as a large gnarled tree, with rough and furrowed bark. In particularly favourable areas of deep sand, it grows as a tall woodland tree, often in association with low heathland species.

Banksia praemorsa
Found growing on the sandy cliffs around Albany in Western Australia, often associated with *Dryandra formosa*, *Banksia praemorsa* (above) is a sturdy, compact shrub of 2 or 3 metres in height. During winter and spring, the shrubs are a striking sight, covered with long candle-like spikes of a rich purplish-red, the outer tips of each style a bright yellow. Honeyeaters dart and chatter among the flowers, secure in the dense foliage from the strong southerly winds which beat the coast.

Banksia prostrata
The creeping furry stems of *Banksia prostrata* may cover one or two metres, sometimes partially buried beneath the sand. The stiff, lobed leaves stand erect from the stems, shielding the flowers, and reaching up to 40 cm high. Golden-brown flower spikes are borne at the ends of the stems; old cones remain as the stem creeps forward. Found only in Western Australia, it grows on sandheaths from north of Perth to east of Esperance. Its form, particularly its leaf shape, varies within its range.

Eucalypts

Australia's gum trees are synonymous with the bush. There is not a part of our continent which does not have one or several species of eucalypt as a strong note in the landscape.

Belonging to the family Myrtaceae, more than six hundred have been recognised and named as distinct species, and many are known to hybridise. Current botanical research is dividing and separating them still further.

From the giants of the cool, temperate forests in Victoria, Tasmania and south-west Western Australia to the squat, many-stemmed mallees of the drier zones, they all possess the floral structure which led to the generic name of eucalypt. This name derives from two Greek words: *eu* meaning good, and *kalyptos* meaning covered — a reference to the fused petals which form the bud-cap or operculum.

The taller species, which include *Eucalyptus regnans* (the Mountain Ash), *Eucalyptus citriodora* (Lemon-scented Gum) and *Eucalyptus diversicolor* (Karri) are insignificant in their flowering. On the other hand, their trunks are magnificent, shafting upwards through the lower-storey vegetation, the dappled light rippling on their smooth limbs.

In general, it is the lower-statured mallees which possess the outstanding flowers. Perhaps the best-known of these is *Eucalyptus macrocarpa* — Rose of the West. Many of these spectacular species grow only in Western Australia. But there are others, not so well-known, which deserve admiration. *Eucalyptus macrandra* boasts clear yellow, sweetly-perfumed blooms; *E. preissiana* bears its brilliant yellow blossoms in winter; *E. caesia* gracefully weeps, its silver stems clustered with rose-pink flowers; and *E. drummondii* produces dainty, nodding yellow buds which open into dense, creamy flowers.

From the northern region, in which many of the bloodwood eucalypts occur, the brilliant orange of *E. miniata* is a breathtaking sight. A tall, rough-barked tree, it stands high above the cliffs of the Katherine Gorge, its flaming blossoms a glorious statement of colour.

While each is known by its generic name of *Eucalyptus*, together with a specific name, each may also be broadly categorised as a gum, stringybark, ash, mallee, bloodwood, box, peppermint or ironbark, and so on. In each case, a specific characteristic has been used to group similar species, and while it may not be strictly accurate in a botanical sense, this method provides a handy tool for recognition by the layman.

Of the many hundreds of species, only a few eucalypts occur outside Australia, mainly on the islands to the north, including New Guinea and Timor. One species is even found in the Philippines.

A tough, fire-resistant, drought-resistant and adaptable family, the eucalypts are part of our natural heritage, a fundamental note in our landscape, and a resource which should neither be taken for granted nor laid waste to greed.

Eucalyptus leucoxylon
Surprisingly variable in its form and colouring, both of leaf and flower, *Eucalyptus leucoxylon* is found in Victoria, South Australia and New South Wales, occupying coastal, sandy heathland and open woodland sites. The many forms (whether cream or pink flowered) are all highly attractive to honeyeaters and bees, and are favoured by apiarists for their fine honey. The form illustrated, known as *Eucalyptus leucoxylon var. rosea*, grows to about 12 metres. Creamy smooth bark covers its trunk and limbs; fine-textured grey foliage droops from the red stems, and the massed deep pink flowers cover the whole canopy from May to November.

Mountain Ash *Eucalyptus regnans*
Famous for its smooth straight-trunked majesty, the Mountain Ash (opposite) is the world's tallest hardwood tree. Cool mountain forests of *Eucalyptus regnans* are common in Tasmania and Victoria, and in both states it is used extensively for its timber. Many species of ferns, and other moisture and shade-loving plants grow in association with these forests, together providing suitable habitat for possums, gliders, wombats and wallabies, as well as numerous birds. Sherbrooke Forest, in Victoria, is famed for both its trees and its lyrebirds. Eucalypts are often identified by their buds, flowers and fruits; *Eucalyptus regnans* buds are small and clustered, the flowers cream, and the little gumnuts, or fruit, shaped like minute tops.

Gungurru *Eucalyptus caesia*
Much has been written about this most graceful small tree of Western Australia (above, left). But while it is frequently given pride of place in many Australian gardens, few people have seen it growing in its natural habitat. In one such area, north of Southern Cross, there are vast red granite rocks, common enough in that region. In a hollow of one of these monoliths, growing in a few centimetres of damp gritty sand, is a community of *Eucalyptus caesia*. So ancient are these specimens that, having reached their vertical growth limits, their slender white branches have swept down to the granite floor and, in some cases, have begun an upward and outward curve again.

Rose of the West *Eucalyptus macrocarpa*
Rose of the West and Mottlecah are both common names for this superbly-flowering species from Western Australia. Found on deep sandy heaths of the wheatbelt country north and east of Perth, it grows as an open, low spreading mallee. Large silver-grey leaves crowd along the mealy stems, and every so often a fat, silver-topped bud peeps out from its protective shelter. As the bud swells, the cap splits from the base, slowly lifting off as the bright stamens stretch and spread apart. The cap may remain in place for a few days, before dropping off; the ground below is always littered with these discarded 'pixie hoods'. The flower matures, its bright beacon attracting willing honeyeaters to pollinate it as they feed. Then, as the stamens wither, the broad ridged and flattened gumnut begins to form, finally producing the largest fruit capsule of all the species.

Eucalyptus macrandra

As attractive as a garden specimen as in its natural coastal habitat of southern Western Australia, *Eucalyptus macrandra* grows as a spreading dense shrub, or a lightly-trunked small tree. Glossy and firm-textured, its deep grey-green leaves are a lovely foil for the bright lemon flowers, which fill the summer air with their perfume. The finger-like bunches of buds are fused to each stem. As the bud matures and swells, the stamens push the cap off, springing out in fluffy bunches to draw insects and honeyeaters in profusion.

Bell-fruited Mallee *Eucalyptus preissiana*

One of Western Australia's many distinctive mallees, the Bell-fruited Mallee (opposite) is a favourite, commonly found on rocky hills and heathlands in the Stirling Range-Esperance region. Sturdy stems, mostly reaching to a height of only 2 metres, but spreading 3 or 4 metres, are densely covered with leathery grey rounded leaves. Fat reddish buds, grouped in threes, perch atop the foliage. Their caps split off in spring, to reveal thick butter-coloured stamens closely massed around the central style. Immediately, honeyeaters make their day-long visits, chattering and flitting from flower to flower, pollinating busily as they feed on the copious nectar. The flowers are followed by intricately-formed bell-shaped fruit, each tightly enclosing its seed until conditions are right for its dispersal.

Grevilleas

Grevillea sericea
Never without flower, this pretty shrub is widespread in the sandstone woodlands of New South Wales, mingling with boronias, eriostemons, ferns and orchids. Growing to 2 metres, with an erect habit, its flowers range from deep pink to mauve and white; the leaves are short and simple. The floral structure of the genus *Grevillea* is actually the same in each species, despite the considerable variation in size, arrangement and colour. The structure differs from that of an open-petalled flower such as a boronia, in which the stamens are free and the petals distinct and separate. In a grevillea, the floral parts are combined in a tepal, which splits open and curls back, revealing the anthers and releasing the long style. Many grevilleas (as with so much of the Proteaceae family) are designed to be pollinated by birds. Flicking its tongue into the nectary at the base of each flower, the honey-eater must touch both style and anther. Thus, at the same time, the bird both collects and transfers pollen.

There are few habitats in which grevilleas cannot be found, and within the several hundred species is an unsurpassed diversity of foliage, flower arrangement and colour.

Named after Charles Greville, a sponsor of botanists, grevilleas belong to the same family — Proteaceae — as such well-known plants as banksias, dryandras and isopogons, and on the scale of evolution are most closely related to hakeas.

It is thought that the beginnings of the genus *Grevillea* were in early rainforests, where some species still grow. While some, such as *Grevillea robusta*, the Silky Oak — much sought for furniture timber — and *G. hilliana*, are large trees, the majority of the species are low-growing or medium-sized shrubs. Many of these smaller species grow no higher than a few centimetres but spread several metres over the ground, forming carpets of interesting texture and colour.

The greatest concentration of species is found in heathlands, and in association with sandstone, such as that around Sydney, in the Grampians in Victoria and the Stirling Range in Western Australia. In these habitats, they are commonly found with other members of their family, together with acacias, epacrids, boronias, correas, melaleucas and eriostemons.

The greatest number of species occurs in Western Australia. The eastern states together share almost as many, while Tasmania has only one, *Grevillea australis*. This species occurs in Victoria and New South Wales as well, occupying open sites in the highest alpine reaches, while in Tasmania, it can be found from sea level to the mountains.

Grevilleas are not generally noted for their perfume, but *G. australis*, and others with similar white or cream flowers, often do possess a strong, sometimes overpowering aroma, which attracts moths, ants, beetles and even flies. These insects, rather than birds, act as pollinators.

Many grevilleas have leaves which are simple in shape, being linear, lanceolate or oval, with no serrations or lobing. There are others, however, with just about every shape and design imaginable. Variations of a holly-leaf shape are numerous, and feathery pinnate structures are equally common.

While not generally as prickly as many of the hakeas, from which grevilleas can be distinguished by their thin seed capsules (hakeas having thick, woody ones), some species are prickly, and provide perfect feeding and nesting habitats for small birds.

As with other members of the Proteaceae family, there is much evidence to suggest parallel evolutionary patterns with many nectar-eating birds, which act as pollinating agents. It is this affinity with our native birds which, in addition to the hardiness of many grevilleas, has resulted in their popularity as garden plants.

The brightness and arrangement of many of the flower spikes, or racemes, is such that they can be seen from great distances, and thus attract pollinators easily. Birds and insects

Grevillea bipinnatifida

A bold shrub of the Darling Ranges south of Perth, *Grevillea bipinnatifida* (above) grows wherever the soil is deep and moist. It may reach a metre or so in both height and width, its foliage much divided and often prickly. Leaf colour varies considerably, with both light green and blue-grey tones being common. The flowers are magnificent; long, waxen racemes of vivid pink, red and orange hang from the branches, and the shrubs are rarely without a bloom.

Grevillea buxifolia

Grey is an unusual colour for any flower, yet New South Wales can boast three grevilleas, all of a woolly silver-grey, and all closely related. *Grevillea buxifolia*, with its box-like leaves, is covered in spring with compact heads of flowers, each invested with a dense velvety fur, and sometimes with a pink undertone. Often found in association with *Grevillea sericea*, it is common among the sandstone forests north of Sydney.

collect pollen on their heads and bodies as they delve into the tubular flowers for nectar. As they seek nectar in the next flower, pollen is transferred and collected again, to achieve a distribution of genes and fertilising of each flower.

The floral structure of grevilleas is quite distinctive. All parts are fused into a tube, which is often thicker at the bottom, and curved and rolled back at the top. The style on which the pollen is deposited is initially hooked, but springs open to protrude from the floral tube.

The arrangement of flowers is as interesting as the diversity of foliage, although one of the most commonly seen is the 'toothbrush flower', in which the styles are arranged in symmetrical formation along the base stem. Spider-flower is the common name for *Grevillea speciosa* in particular, but refers to all those species with the flowers arranged like wheels, or outspread spider's legs.

Almost all grevillea flowers combine several colours, making close investigation, particularly with a hand-lens, most rewarding. Pink and green, mauve and cream, black and green, orange and red, claret and gold, pink and orange, translucent blue-greens and woolly greys can all be seen.

Only a few species of *Grevillea* occur outside Australia, and the genus is considered one of the major elements of the flora of this continent.

Grevillea juniperina
A New South Wales and ACT species, *Grevillea juniperina* (left) varies considerably in flower colour and habit. Red, orange or yellow forms, together with bushy, low or prostrate habits, make this an interesting species. The form pictured is yellow flowered and prostrate, covering 2 metres as a dense carpet, generally in open forest on shaly soils. Leaves are rigid, short and pointed, the stems covered with fine white down, and the profusion of flowers are at their best in spring time.

Silky Oak   *Grevillea robusta*
Tall tree of Queensland rainforest and descendant of some of the earliest Proteaceae plants, *Grevillea robusta* or Silky Oak (lower left) is a mass of fiery gold in late spring. Widely planted as a garden, parkland and street tree, it has a dark, furrowed trunk and a spreading canopy of finely divided leaves. The racemes of flower, 10 cm or more in length, are arrayed in 'toothbrush' formation, spraying out either side of the stem, and glistening with nectar for bird and insect.

Grevillea victoriae
There are several distinct populations of this species (opposite) which is known only from Victoria and New South Wales. Generally alpine and sub-alpine in its distribution (a rare prostrate form occurs on Sentinel Peak, a pinnacle in the Kosciusko region), it varies in leaf size, flower colour and habit. The form pictured, a sturdy shrub of 2 to 3 metres, grows in a small pocket of shady forest in Victoria's east Gippsland, in high country, but well below the snowline. Its apple-green leaves are covered with fine hairs, and the large clusters of flowers are unusually bright. The more typical form of *Grevillea victoriae* has grey-green leaves, distinctly veined, and rusty flowers; it is common among rocky, lightly wooded areas above the snowline.

The High Country

Perhaps some of the most inspiring of Australia's many faces are to be found in the regions above 1500 metres — the country known as the Alps.

The three main alpine regions occur in Tasmania, Victoria and New South Wales, each possessing its own characteristic landforms and flora. The combined area of snow country totals only 0.15 per cent of Australia — with the Snowy Mountains one of the most extensive areas.

Tasmania blends a wilderness of rugged spiked peaks, limpid tarns, sheer rock faces, sparkling quartzite, and strange and beautiful plants such as the endemic Deciduous Beech (*Nothofagus gunnii*).

Victoria's high country is somewhat gentler in aspect, more curved and undulating in nature, with the expanse of the Bogong High Plains the cradle of its beauty. All around are the forested slopes of misty ranges, each climbing to a summit clothed with flowers and smooth-trunked snowgums.

Australia's highest mountain, Mt Kosciusko, is part of the Main Range of the Snowy Mountains, which form the bulk of New South Wales' alpine lands. With range after rolling range, drained by myriad streams, the grandeur of the Great Dividing Range is nowhere more clearly seen than in this dramatic vista.

To walk over this high country is to experience a 'one-ness' with nature: an identification with the beginnings of life. It has something to do with the clarity, the purity of the air; something to do with the utter silence; with the wind, the ripple of grass, the field upon field of flowers, and with the last of the snowdrifts, or the silent swirl of the first snow settling softly, gently over the earth.

One cannot feel alienated here. Rather, there is an aura of quiet acceptance, for the person who comes to know something of the essence of the high country at the same time knows a little more of himself. It draws one again and again, for within its enfolding, gentle silence one experiences release.

The intense extremes of cold of the high country have produced a flora which is tough. Snow covers much of the region for at least six months each year, and even the summer months bring forth wild storms and sharp drops in temperature. Yet much of the flora is delicate, and would seem almost too frail to withstand the rigours of the climate.

One of the most fragile is *Caltha introloba*, a tiny 'buttercup' of purest white. Impatient for release from the deep blanket of winter's snow, at the first signs of thaw its exquisite flowers appear far back in the caverns formed by the melt-water of brooks and streams. Firmly rooted among the gravel and stones of the stream bed, its blooms are held above the shallow rippling water — rarely catching a glimpse of the sun or sky.

Pollinated quickly, by the time the snow cavern edge has receded to expose them fully, the flowers are replaced by seed-heads, nestling among dense tufts of pointed, lobed leaves. Thus, before summer is fully upon the high country, one of its

Craspedia species
Golden heads of *Craspedia species* carpet a gently sloping herbfield. One of the joys of walking in the alps is to chance upon such a field of waving colour, for the season is brief, and perhaps in a few days the peak of flowering will have faded. Mt Feathertop, Victoria's second highest peak, lies sun-bathed in the distance.

Trigger Plant *Stylidium graminifolium*
Found from sea level to the highest alpine herbfields, this particular Trigger Plant (opposite) exhibits remarkable environmental adaptability. While colours of pale pink to almost white prevail at lower altitudes, particularly in open forests, the alpine form is typically a deep rosy pink to magenta. Drifts of these dainty heads may be seen from some distance, colouring the springy heathland, and attracting moths and butterflies to alight upon the tiny flowers. A miracle of design is thus demonstrated, for the sensitive pollen-presenter is activated by the insect's foraging, and a speck of pollen deposited on its body is exchanged for another and another on subsequent flowers.

Kunzea ericifolia (syn *K. muelleri*)
Like most members of the family Myrtaceae, this *Kunzea* (above, below) has delightfully aromatic foliage. Soft, grey-green or sometimes silvery, its tiny leaves may be almost hidden beneath the massed fluffy yellow flowers during mid-summer. At high altitudes, in the Kosciusko region, it may be seen on steep mountain faces as a pure community. In lower areas of New South Wales and Victoria, among more diverse associations, it will be found as a rounded clump, or perhaps adapted to the contours of a boulder, from which it receives warmth and protection.

numerous species has ensured its place in the future, with several months of warmth and light in which to germinate and establish new little plants.

A close relationship is apparent between the alpine flora of Australia and New Zealand. Before the drifting apart of the great southern land mass, much evolution of the flora had occurred, together with interchange of species between areas. But further speciation developed with the isolation of the continents, and new niches and habitats were created as the landform and climate altered.

Dense forests clothe the flanks of the mountains to the treeline — the point at which trees cease to occur as part of the flora. The treeline is generally regarded as being at about the point of mean mid-summer temperatures of 10°C. In the Kosciusko region, this is about 1830 metres above sea level. Above that, to the summit at 2228 metres, the flora is regarded as truly alpine.

Apart from the copses of gnarled, twisted snowgums which are found in sheltered gullies and on hillsides, the dominant, characteristic flora in all this high country is heathland. In a botanical sense, this flora can be divided into far more complex groups, to include feldmarks, fens and bogs, and heathland and herbfield of varying categories. The edges of lakes, rocky outcrops and rock screes also provide for specialised groups of plants.

It is a flora which deserves the utmost attention. Its exquisite detail invites the closest inspection, on hands and knees — perhaps with a hand-lens as well — to reveal its full beauty and wonder.

Stackhousia pulvinaris
A sweet fragrance wafts strongly in the breeze. Where is it coming from? There, perhaps ten metres away, is a patch of cream on the earth, but surely such a perfume would not come from those minute flowers. But it does. Exquisite, tiny upturned stars lie closely matted over a bed of compact little leaves, flowering throughout the brief summer months, and found over all the broad alpine plains.

Ranunculus anemoneus
Surely the loveliest of all Australia's alpine flowers, the purity of this snow-loving species (right) is breathtaking. Like its small companion, *Caltha introloba*, it dwells in the receding snows, often in a cascade of melt-water or far back in a hollowed cavern, where the light is dim and the air chill. The largest of all our *Ranunculus* species, it belongs on the highest slopes of the Kosciusko area; a treasure to be sought and a joy to find. Robust dark green foliage guards the strong stems, often forming a frill beneath the cluster of many-petalled flowers.

Snow Daisies *Celmisia asteliifolia*
The silver carpets which clothe the slopes, valleys and broad tops of both the Kosciusko region and the Bogong country are composed of countless millions of these Snow Daisies. Where it is exposed to severe wind, *Celmisia* will grow as a tiny, almost flat tuft of leaves, with two or three short little flower stems. In sheltered sites it grows as a graceful, long-stemmed herb, its large snowy (or occasionally pink) petals perfectly formed and of silken texture.

Marsh Marigold *Caltha introloba*
Exquisite companion of the melting snow, this small flower is among the first to bud, bloom and set its seed as the alpine summer begins. Starry white, cream or palest pink flowers nestle among bright green twin-lobed leaves, and will be seen in the shallow icy rills below snow drifts, or deep inside the cool ice caverns which form over the swift, gravel-bottomed creeks.

Helichrysum acuminatum
Glowing under clear alpine skies, the burnished gold of these massed everlastings is an unforgettable sight. Only a few centimetres high, they drift over summit and hollow, so dense that one cannot step between the flowers. Tasmania, Victoria and New South Wales share this lovely herb, which appears faithfully each summer to add its layer of sunshine to every alpine meadow.

Bogong Daisy-bush *Olearia frostii*
This woolly shrublet is endemic to Victoria's high plains, where it may often be found among the dense tussocks of snow grass. The pale mauve daisy flowers are 2-3 cm across, and combine with the silvery, soft leaves to make a charming picture.

The last of the snow lies amid the flowering sunny meadows of a Kosciusko summer. Blue, gold, silver, mauve and pink blend, drift and sway; the air is piercingly silent and intoxicatingly fragrant, throbbing with the deep and steady pulse of new life.

Neopaxia australasica
Snow-white, cream or pink, these little flowers express the simplicity and natural harmony of the alpine atmosphere. Sometimes growing in thick carpets several metres across, and flowering so densely that little green can be seen, they also occur in pockets of just a few flowers, embedded in damp sphagnum moss. Apparently frail, *Neopaxia* serves a most important function in its environment; its creeping, matting root system stabilises the soil and controls erosion. Slightly succulent, its fleshy leaves and stems mark it as a member of the Portulacaeae family. Its distribution is considerable, from alpine to coast, and in most states as well as New Zealand.

Ranunculus graniticola, Scleranthus biflorus,
and *Hovea purpurea*

In this little alpine garden, the smooth mossy *Scleranthus biflorus* spreads its emerald cushion gently amongst golden *Ranunculus graniticola* and purple *Hovea purpurea*. The *Hovea* flowers early in the high country, turning the slopes and hills to misty shades of blue and mauve. The many species of *Ranunculus* linger to late summer, their seed setting with just enough time to fall before the first snows blanket the earth.

The Shoreline

Small Stilted Mangrove *Rhizophora stylosa*
Twenty-nine species of mangrove are known around Australia's coast, with most occurring in the tropical zone on protected low tidal mud-flats and estuaries. The species pictured shows the stilt or prop roots associated with this particular genus. These aerating roots, known as pneumatophores, not only support the tree, but absorb oxygen through their tissues during low tide. Aerial support roots descending from the lower branches can also be seen. The most widespread species, *Avicennia marina* (the Grey Mangrove) employs a different method of root aeration, in which forests of little pneumatophores project vertically above the surface at low tide to absorb oxygen, which is then used by the tree during inundation. The mangrove community is extremely important ecologically, particularly in its vital role as a nursery and feeding ground for a wide range of marine life.

The range of flora found around the entire seaboard of Australia varies quite surprisingly. Soil and rock structure, aspect and prevailing winds, together with the overall climatic pattern of the region, combine to create differing environments that may contain just one dominant plant species, an association of several key species, or a whole complex of many species.

The northern and more tropical regions typically support mangrove communities. Visually these may not at first appear inspiring. Ecologically, they are vitally important, providing safe nurseries for marine life. However, not all coastal northern Australia is bordered with mangroves. In many places along the eastern coast, the rainforest fronts right to the edge of the beach, indicating a strong tolerance of salt, wind and spray. The northern beaches of Western Australia often run back into vast plains of heathland, spinifex and tussock, or low windswept saltbush communities, punctuated occasionally by stunted mallee.

The southern half of the continent's coastline supports vast and beautiful stretches of heathland. The coastal heathlands, of Western Australia in particular, contain unbounded variety, and no matter what time of year, there will be much to see in flower. In mid-winter, only metres from the pounding seas and spray of the Southern Ocean near Esperance, the vivid orange and red velvet flowers of *Lechenaultia formosa*, and the furry yellow tubes of species of conostylis nestle among granite chips and boulders. In mid-summer, the plunging cliff faces of the Mid Mt Barrens, east of Albany, are scarlet with the brushes of *Regelia velutina*, a silver-leaved shrub known only from this region.

And spring in any region of heathland, anywhere in the whole continent, brings spectacular displays of wildflowers.

The coastal heathlands of the eastern states, while they contain a smaller proportion of species, are as full of interest as their western counterparts. In combination with these heathlands are the forests which clothe the hillsides, marking the regions of moist depressions or creeksides with taller growth and changes in species. Where quieter, more sheltered habitats occur in estuaries and inlets, the forest may become more dominant. Hillsides dropping into the depths of the Hawkesbury River, Mallacoota Inlet or Macquarie Harbour are dense with tall trees and a varied understorey of shrubs and creepers.

Sand dunes provide a habitat for a more limited but nevertheless interesting range of flora, often confined to a mere handful of species. The Coast Wattle, *Acacia sophorae*, dominates in such situations on the east coast, often forming impenetrable thickets many metres wide, and three to four metres high, in the hollows between the dunes. *Banksia integrifolia* may also be found, often in association with the wattle, thus forming an outstanding habitat for birds.

Great granite sheets and boulders tumble into the sea, producing a shoreline of spectacular beauty. The Freycinet Peninsula, which juts out from the east coast of Tasmania, owes its present form to the flooding of Tasmania's coast, which followed the melting of the last ice age. Coastal heathland combines with open forest; grasstrees, casuarinas and acacias cast dappled shade over orchids and other small wildflowers. While much of the granite is naturally a red-brown tone, a closer look (right) will reveal that some of the colour is due to lichen. These early forms of plant life colonise not only rock, but also soil and other plants, and exist in almost every part of the world. Carpets of gold, rust, yellow and red, green, grey or white will be found, slowly spreading in patches across the rock, and plainly able to exist in the salty environment. Viewed through a hand lens, they are an intriguing world in miniature.

Communities of grasses, herbs and succulents occur in recognisable zones along the foreshore, and the roots and spreading, flattened branches of all the dune species serve a vital purpose in stabilising the entire dune structure. Wind erosion is severe on exposed surfaces, and rolling hills of shifting sand may be seen in the majority of sand dune regions.

Though varied from state to state, the coastal flora of the entire continent exhibits several common characteristics. It is tolerant of soils which are low in humus and which dry out readily in summer; and it is resistant to fierce, salt-laden winds and the drenching spray from the ocean itself. Temperatures can range from bitter cold to dessicating heat, and still the plants survive — evidence of another characteristic: their remarkable adaptability.

Albany Bottlebrush *Callistemon speciosus*
The Albany Bottlebrush lifts its heavy blooms high above the rigid foliage, blazing beacons to ever-willing honeyeaters. The deep coastal sands and swamp margins of the Albany district are the main habitats for this species, one of the only two found in Western Australia. Callistemons are more widespread in eastern Australia, but almost without exception, their favoured habitats include plenty of moisture, either in swamps or on stream-sides, and ample sunshine.

Lechenaultia formosa
Growing only a few metres from the surging seas of the Southern Ocean, on the granite headlands of Cape Le Grand National Park in Western Australia, *Lechenaultia formosa* bears its scarlet velvet flowers through most of winter. Such steep rocky slopes are the habitat for an extraordinary range of small, apparently delicate plants, which have adapted to conditions of fierce wind, heavy salt air and sea spray. Their low, often prostrate, growth habits offer minimal resistance to the wind, and their tenacious roots are able to grip firmly amongst the rock crevices, holding them snug and secure.

Variable Groundsel *Senecio lautus*

Aptly named, this small creeping herb will be found in a wide range of habitats from coastal to alpine, and in every state. Leaf shape and number of flower heads vary considerably, and while the flowers differ a little, they are always a bright cheerful yellow. As a coastal plant, it may be seen scrambling and carpeting the sheltered hollows among sand dunes, or as a flat mat on exposed faces and windy headlands.

Much of Australia's coastline is rugged, beautiful and, where settlement and tourism have not reached out, retains a wild grandeur of its own. Apart from widely separated settlements, the long sweep of indented coast between Augusta and Esperance contains much for the naturalist, whatever his special interests may be. Vast heathlands, crammed with rare and lovely plants, stretch on either hand, while the Southern Ocean rolls in upon long white beaches, or crashes up against jagged headlands. Reserves protect stretches of coast, in particular the magnificent Barren Ranges within the Fitzgerald River National Park, which lies between Albany and Hopetoun.

Cushion Bush and Pigface
Calocephalus brownii and *Carpobrotus rossii*
The compact silvery ball of the Cushion Bush, *Calocephalus brownii*, is a familiar sight along much of the coast, in all states but Queensland. Growing up to a metre in height, and often much wider, it will be found well to the fore along the beach front. Its dense branchlets are tightly tangled, and during spring and summer are dotted profusely with small cream button-like flowers. Carpeting the sand is the fleshy-leaved Pigface, *Carpobrotus rossii*, its bright pink flowers several centimetres across. The succulent leaves contain a great deal of moisture, enabling the plant to withstand the hot, dry conditions which often prevail along the coast.

Wildflower Forests

Forests clothe most of Australia's eastern freeboard, varying in density, height and species range according to climate, soil and topography. Tall, broad-leaved rainforest species of the northern coasts allow little light to penetrate the canopy, and the understorey is largely composed of ferns and species which have adapted to a dark, moist environment. Southern rainforests are found in remnant pockets in Victoria, and in Tasmania, where Beech and Sassafras are more dominant.

Rainforest gives way in cooler or drier conditions to the hard-leaved, wet and dry sclerophyll forests which cover most of the Great Dividing Range, and are also found in the southern corner of Western Australia.

The cool, wet sclerophyll forests are typified by tall eucalypts such as *E. globulus* in Tasmania, *E. regnans* in Victoria, and *E. diversicolor* in Western Australia, together with an understorey of ferns and soft-leaved shrubs. In the dry sclerophyll forests, the trees are more widely spaced and the canopy more open. Light penetrates to the forest floor, allowing a wide range of small and fine-leaved species to grow.

During summer, autumn and winter, there are few wildflowers to be seen in these drier forested regions of Australia. However, during spring the forest floor is crowded with what, to many people, are the true wildflowers of the bush. It is their daintiness and apparent fragility which is so appealing. Small lilies such as *Burchardia umbellata*, the creamy Milkmaids, or the sky-blue *Chamaescilla corymbosa* (Blue Squill) dot the carpet of leaf litter, twigs and gumnuts. The drifting sweet fragrance of *Dichopogon strictus* (the Chocolate Lily) mingles with the heady blossom-scent of the eucalypts and acacias.

A vast array of terrestrial orchids nestle shyly beneath trees. Where you find one orchid, with a little searching, you will spy several more. Greenhood is the charming and appropriate common name given to the large group of *Pterostylis* species. Borne singly on sturdy stems, or grouped along the stem, the flowers are readily recognised by their translucent, curved hoods, generally green but occasionally reddish-brown, or perhaps striped with cream.

Most of the terrestrial orchids have appealing and appropriate common names. Parson's Bands describes the two long white sepals of *Eriochilus cucullata*. Pink Fingers refers to the spreading, delicate pink floral segments of *Caladenia carnea*. Flying Duck is the whimsical and descriptive name for *Caleana major*; and the Beard Orchid, *Calochilus robertsonii*, is immediately recognised by its long woolly labellum. Spider, sun, leek, donkey, bird, helmet, waxlip, cinnamon bells and hyacinth are other common names which immediately focus attention on the appropriate features of the species to which they refer.

Bird Orchid *Chiloglottis gunnii*
Quaintly appealing, this little orchid (opposite) nestles close to the ground, sometimes half hidden among the leaf litter and moss which accumulates in shady, moist patches of the forest. Its open-mouthed, upturned face is the reason for its common name; the resemblance to a baby bird begging for food is amusingly close. Widespread in the forests of Victoria, New South Wales and Tasmania, and flowering in spring, it is one of seven species of the genus *Chiloglottis*, which is confined to eastern Australia. The family Orchidaceae, to which it belongs, is world-wide in its distribution. In Australia alone, about 750 species are recorded, most of them endemic (that is, occurring nowhere else).

Brown Beard-orchid *Calochilus robertsonii*
The Brown Beard-orchid is often found in eastern Australian forests, tucked at the foot of stringybarks, half hidden amongst tussock-grass, rambling buttercups and moss. It will also be found in heathland, and is recorded for all states. Several flowers are borne on one smooth pale green stem; as each opens, the long labellum, or 'lip' protrudes to show its distinctive brown woolly beard. There are ten species in the genus *Calochilus*, some of which occur in New Zealand and New Caledonia.

Tufts of *Brunonia australis*, the Blue Pincushions, lay a film of soft colour through the forests. The yellow faces of many goodenia species reflect the filtered sunlight; and the vivid purple of *Hardenbergia violacea* in the eastern states, and *H. comptoniana* in the west, is a rich addition to the forest's glory. Scrambling with the western species will be the glowing pink, orange and red pea-flowers of *Kennedia coccinea*, the two combining thickly to drape entire patches of bush.

Egg-and-Bacon is the comical, yet charming epithet bestowed by early settlers upon the many yellow, brown and red pea-flowers of the Fabaceae family. *Dillwynia, Pultenaea, Daviesia, Eutaxia, Platylobium, Gastrolobium* and *Bossiaea* are widespread members. Many will be found in open forests, forming bright clumps of colour, their peak of flowering so intense that hardly a leaf can be seen on the shrub.

Lighting up the grey-green of the forest from mid-winter to early summer are the wattles. As one species fades, another awaits its moment to burst into fluffy perfumed beauty. No area of bush is without them, for their extraordinary adaptability has earned for them a place in every corner of the continent.

Grevilleas, too, will be found in much of Australia's forested regions. They include carpeting species such as red-flowered *G. laurifolia* and *G. repens*; softly-arching shrubs of the highly variable *G. alpina* group, in golds, reds, orange, pinks and white; compact clumps of *G. wilsonii*, its cherry-red waxy flowers so prominent in the forests of the Darling Range near Perth; or taller shrubs, such as *G. rosmarinifolia, G. shiressii, G. victoriae*, and the rare *G. longistyla*, known from isolated pockets in New South Wales and southern Queensland.

The Rutaceae family is strongly represented in the drier forests. Boronias, correas, croweas, eriostemons, phebaliums, zierias and asterolasias contain many individual species which are noted for their dainty red, pink, white or yellow, starry or bell-shaped flowers. Almost all possess a strongly pungent fragrance to their foliage, and certain boronias, notably the famous brown *B. megastigma*, are immediately associated with a unique fragrance of flower, which has led to their wide appreciation as cut flowers.

Hibbertias, which belong to the family Dilleniaceae, are among the sunniest and most cheerful of the forest wild flowers across the continent. Their open-petalled flowers spangle climber, scrambler, clump and shrub with reckless gaiety. Their hardiness has won them a place in many gardens, and their flowers of pure yellow, gold and orange contrast brilliantly with their muted grey and green foliage.

The nodding pink bells of many species of tetratheca cluster closely along their slender, arching stems — each plant a small cloud of colour. *Dianella revoluta* forms dense clumps, its dark, stout, strap-like leaves surrounding the tall stalks of blue and yellow flowers. Many wahlenbergia species grow profusely, their blue bells upturned on fine, pale stems; and blue dampieras and scaevolas tangle among the native grasses.

Baeckea ramosissima, variable in habit, is commonly found in southern forests, its pale to deep pink open-petalled flowers massed on fine, wiry stems. The *Myrtaceae* family, to which baeckeas belong, contains many forest-loving species, including kunzeas, leptospermums, callistemons, melaleucas, thryptomene, darwinias and micromyrtus. Particularly characteristic of all the family is the fragrance contained in the foliage. Released after rain, or when a leaf is lightly crushed, it has a lingering and satisfying pungency.

The forests contain a wealth of interest and beauty, all of which may be seen, smelt, felt and heard by anyone who takes the time to wander slowly and observe quietly.

Fairy Waxflower *Eriostemon verrucosus*
This species belongs to the family Rutaceae, which includes boronias, croweas, phebaliums and correas. This small shrub of less than a metre is often found in open, dry woodland, particularly in the goldfields district of Victoria, but also in South Australia, New South Wales and Tasmania. Generally it is seen with white to pale pink flowers; occasionally, the buds will be deep rosy-pink, opening to a white flower; the foliage is grey, warty in texture and slightly curled. The form photographed, however, shows a particularly lovely double-petalled flower, each like a miniature water-lily, with delicate orange stamens.

Purple Coral Pea   *Hardenbergia violacea*
True harbinger of spring, the Sarsaparilla or Purple Coral-pea twines through the forests and woodlands of Tasmania, Victoria, New South Wales and Queensland. Sometimes it may cover the ground as a carpet; mostly it is found rambling and twisting about old stumps or along slender stems of a supporting shrub. Rarely is there a patch of dry forest without its purple glory festooning earth, limb or twig. Western Australia has its own species, *Hardenbergia comptoniana*. This, too, loves the dappled forest environment, scrambling through or spreading its mantle over shrubs, bracken and fallen logs.

Common Fringe-lily   *Thysanotus tuberosus*
Widespread in the forests and scrublands in almost every state of the continent, the Common Fringe-lily (right) makes a dainty show in spring and summer. Each petal is delicately fringed with soft filaments, a characteristic of all species of *Thysanotus* which imparts a particular charm. A member of the Liliaceae family, it grows as a slender herb to 30 cm, and the flowers are borne on several fine branchlets, opening fully on sunny days but closing at evening and when skies are overcast.

Small Forest Flowers
The massed flowering which is one of the charms of the drier forests is seen to perfection in such areas as the Grampians, in Western Victoria. Late winter and spring bring forth a profusion of flowers: small heathland plants such as *Epacris impressa* (Pink Heath), *Tetratheca ciliata* (Black-eyed Susan), and *Styphelia strigosa* (Peach Heath) combine their pretty pastel colours with splashes of yellow acacia in the backdrop of forest.

Native Violet *Viola hederacea*
No less charming because it is so common, the Native Violet (left) colonises any damp, shady nook in most forests of South Australia, Tasmania, Victoria, New South Wales and Queensland. Only a few centimetres high, the smooth green ivy-shaped leaves grow thickly, creeping by underground stems wherever the soil is moist. Dainty white and purple flowers are borne on single stems throughout the year.

Egg-and-Bacon *Pultenaea paleacea*
A small, fine-stemmed shrublet, or occasionally a delicate open mat, this fragile species is fairly common in eastern Victoria, and occurs also in New South Wales and Queensland, in open woodland and lowland heaths. It is a mass of flower during spring, but its habit of twining closely amongst other low-growing plants, often in marginally swampy conditions, has kept it from being well-known. Bush-pea, or Egg-and-Bacon, are the common names for this brightly-coloured group, many of which will be found flowering in woodland and heath in each state, throughout the spring months. The pea family (Fabaceae) is found throughout the world, and numerically is second only to the family Orchidaceae.

Guinea-flower *Hibbertia procumbens*
One of many Guinea-flowers, a genus which is widespread throughout Australia, this prostrate matting species (left) is found in moist, partly-shaded conditions in southern Victoria and Tasmania. Buttercup-yellow flowers dot the foliage throughout the summer, sometimes so thickly that little of the dark fine foliage can be seen. Not all hibbertias are yellow, however. Two particularly beautiful species from Western Australia have orange flowers: *Hibbertia stellaris*, which becomes a little ball of multi-hued orange, and *Hibbertia miniata*, a rare low-growing shrub of jarrah forests. Almost all the genus is forest-loving, lighting the undergrowth with a profusion of bright, glowing, open-petalled flowers.

Bush-pea *Pultenaea pedunculata*
A prostrate, matting carpet, this dainty Bush-pea is known to colonise dry, often shaly or rocky soil in open woodland of Victoria, South Australia, New South Wales and Tasmania. As it creeps, it may send down roots from the stems, effectively stabilising soil and checking erosion on banks and steep slopes. Older plants in bushland areas may be found 3 or 4 metres across, the plants sometimes meeting, so that the ground is a solid carpet of dark green, no higher than 6 or 8 cm. Flower colour varies considerably. The most common combination is that of the form pictured: gold and red, but deep pink, pure yellow or gold, and dark red are all known to occur as solid colours. Fine, close foliage covers the stems densely, making it a most attractive plant.

Kangaroo Paw *Anigozanthos manglesii*
Widely-known as Western Australia's floral emblem, the red and green Kangaroo Paw (above, left) flowers profusely through the woodlands of the Darling Ranges south of Perth, as well as extending north on the sandplains as far as Shark Bay. Long felted dark red stems rise above the grey-green strap-like leaves. Tightly closed, the buds resemble paws. As they unfurl, the lower flowers first, each long velvety-green tube splits and curls back, to expose the yellow stamens and anthers. This intriguing structure is then easily pollinated by honeyeaters, which brush their heads against the anthers as they delve their tongues deep into the nectary at the base of the flower, and then unwittingly transfer a dusting of pollen to the next flower.

Spring in the forest
Not a leaf has been touched, not a flower added, in this charming spring carpet in a Western Australian forest.

Sandstone Gardens

Australia's sandstone country holds some of the most extensive and beautiful natural rock gardens in the world. New South Wales, Victoria and Western Australia all contain magnificent examples. Each is quite different in its flora, yet each supports an incredible diversity of wildflowers.

For sheer sculptural beauty, the rock itself is outstanding. Weathered to smooth blocks and sheets, its colour is muted and glowing. Sit on a sun-warmed slab and allow your eyes to wander over shelf and ledge, each narrow crack and cranny offering a foot-hold for some little plant; or climb by natural step and stair to the summit of some massive bluff, tip-tilted millions of years ago by ancient subterranean upheavals.

Sydney Rock Rose *Boronia serrulata*
The Sydney Rock Rose (opposite), together with Western Australia's Brown Boronia (*B. megastigma*) are the two best-known of the many members of this genus, which belongs to the family Rutaceae. Many boronias are perfumed in both flower and foliage, though none is quite so fragrant as *Boronia megastigma*. Growing only in the sandstone region between Gosford and Royal National Park, *Boronia serrulata* is inconspicuous unless in flower. It nestles shyly amongst tangled low heathland plants, often in shady cool hollows, or on the ledges of the honey-coloured sandstone where moisture may collect. But in spring, when its rosy four-petalled flowers bloom for many weeks, it may be spied from a distance. The flowers cluster thickly along each stem, which is sheathed in smooth, flattened leaves.

Flannel Flowers *Actinotus helianthii*
Snipped from cream velvet, the Flannel Flower must hold pride of place as one of Australia's most beautiful wildflowers. It grows in sandstone woodlands where it is a slender open shrub, and on sandy coastal heaths from the south coast of New South Wales to south-eastern Queensland, where it will be seen as a compact, low shrub with strong woody stems. The foliage is soft, silvery grey and much divided. Flowering prolifically from early spring to almost the end of summer, each of the perfect blooms is actually a dense globular head of minute flowers, surrounded by soft felty bracts which look like petals. Though one might expect it to be related to the daisy or Asteraceae family, in fact it belongs to the family Umbelliferae, which contains several little-known genera, quite a number of which are found in alpine regions.

The Hawkesbury Sandstone

Fuchsia Heath *Epacris longiflora*
Beloved of honeyeaters, particularly Eastern Spinebills, this member of the heath (Epacridaceae) family is a strong component of the Hawkesbury sandstone, common on ledges and cliff faces, particularly where moisture collects. Flowering almost all year round, the wiry stems arch and tangle over, through and around shrub and rock. The bells hang in their serried rows, as neat as a pin, their white tips contrasting freshly with the scarlet and green. It is confined to sandstone, in particular that around Sydney.

North of Sydney, within a half-hour drive of the city, the scenery is still very much as it must have appeared to the early explorers. The very nature of the Sydney sandstone confines suburban sprawl and holiday settlements to the shoreline, and the tops of the flattened ridges which separate the valleys.

The deep and serene Hawkesbury River lies basking in the warmth of spring and summer, its many inlets inviting exploration, and its tributaries tinkling and cascading over rock walls, worn shelves, and between weathered boulders in shaded gullies, to meet the green salt water of the river. The lower Hawkesbury is, in fact, a vast network of valleys drowned by the sea, and the tidal influence carries many kilometres up the river's length.

To wander on foot through this ever-changing garden is to meet, on every hand, a new plant, a new vista. To many, the Hawkesbury sandstone is epitomised by its characteristic Applegums, *Angophora costata*. Apricot-skinned and smooth, their trunks curve and blend with the ancient rock, wrinkling and bulging against pressure, often leaning at fantastic angles, their upstretched limbs bearing a canopy of fresh green foliage. Nestling beneath, amongst the litter of bark curls, twigs and leaves, are flannel flowers, rock orchids, eriostemons, boronias and croweas, heaths and hibbertias.

The woolly grey flowers of *Grevillea buxifolia*, the soft pink-mauve of *G. sericea* and the bright scarlet spiders of *G. speciosa* are often seen in combination. An occasional Waratah, *Telopea speciosissima*, may be found, its brilliant scarlet bracts enclosing compact bundles of curled flowers. *Eriostemon australasius*, one of the largest and most beautiful of the wax-flowers, is covered with shell-pink star-flowers in spring, and at knee-height one may brush through thickets of the deep pink Sydney Rock Rose, *Boronia serrulata*, releasing the strong fragrance from the flattened leaves.

Graceful and fine-foliaged, *Persoonia pinifolia* begins flowering in mid-summer, and its tubular yellow flowers, clustered in pendant bundles, are still opening well into autumn. Many heath plants colonise the moisture-holding crevices in the rock shelves and sheer walls. Much loved by the Eastern Spinebills is the Fuchsia Heath, *Epacris longiflora*. Slender wiry stems arch out, each bearing a symmetrical array of long scarlet, white-tipped bells, full of nectar, and flowering for most of the year. *Epacris reclinata* clings to rocky faces, its deep rose-pink bells clustered thickly along its erect stems.

Taking advantage of the moisture-laden rock, and found on almost every shelf, is *Banksia ericifolia*. Hundreds of deep gold to red cones cover these shrubs during winter and spring, and the constant call and flight of feeding honeyeaters is a musical testimony to the quantities of nectar produced. In swampy patches, amid the thickets of rushes and reeds, flashes of scarlet tell of *Blandfordia nobilis*, the Christmas Bells of New South Wales.

Christmas Bells
Blandfordia nobilis and *Blandfordia grandiflora*

Christmas Bells, *Blandfordia nobilis* (above), are well-loved by the people of New South Wales. They do indeed flower at Christmas, and occasionally one may come across a swampy patch entirely covered with them. Growing only 30 to 40 cm high, each stem bears a cluster of nodding bells 2 to 3 centimetres long. Waxy and firm in texture, they glow brilliantly in the sunshine. Heathlands and swamps are their chosen habitats around the Hawkesbury River. North of the Hawkesbury, in the Gosford area, one may also find the dramatic large-flowered *Blandfordia grandiflora* (right). The colouring is similar, but the size of each bell more than justifies its specific name. Blandfordias belong to the family Liliaceae, which in Australia contains a large number of unusual members, quite different from the conventional concept of lilies.

Persoonia pinifolia

Softly weeping amid the dappled shadows of the sandstone forest, *Persoonia pinifolia* (above, right) spills its golden flowers in graceful cascades over its fine foliage and arching branches. Flowering is followed by equally attractive strings of bronze-green fruit, known as drupes. *Persoonia* is a large and variable genus of the Proteaceae family, with species found on both sides of the continent. All bear tubular yellow flowers which, as they mature, split and curl back, freeing the stamens and style. Many persoonias develop particularly attractive bark. Tissue-thin, it peels back in layers to reveal tones of deepest red and tan.

Waxflower *Eriostemon australasius*

Though not common, this species (opposite) is New South Wales' best known Waxflower, and occurs also in southern Queensland, while a broad-leaved form is known from northern Queensland. It grows in the shady sandstone woodlands as an erect shrub of about 2 metres. The grey-green leathery leaves are an attractive foil for the mass of large, shell-pink waxy blooms which cover the bush throughout the spring. The family Rutaceae, to which eriostemons belong, is world-wide in its distribution, and in Australia contains some of our best-loved small shrubby wildflowers — croweas, correas, phebaliums and boronias.

The Grampians

The Grampians rise abruptly from the plains of Western Victoria, terminating the Great Dividing Range in some of the state's most spectacular scenery. Four roughly parallel ranges are separated by wide forested valleys; the precipitous eastern scarps are backed by gentle western slopes, typified in this view of Redman Bluff, part of the William Range. Faulting, uplift and weathering of softer surrounding rock have left the Grampians standing as high ranges, once surrounded by ancient seas.

Grass-trees *Xanthorrhoea australis*
Spectacular flowering of the Grass-trees, often induced by bushfires, produces an almost primeval sight. Each spike is covered with thousands of minute white flowers, glistening with nectar for hungry bees and honeyeaters. The long skirts of grassy foliage may reach to the ground on unburnt specimens, completely hiding the thick sturdy black trunks. Remarkable for its longevity and for its slow rate of growth, this species, a member of the family Xanthorrhoeaceae, is also found in Tasmania and South Australia.

The main Victorian contribution to sandstone flora is found in the Grampians, a region of rocky escarpments and precipitous bluffs almost twice as old as the sandstone of New South Wales.

Late winter and spring bring the peak of flowering to these forests, heathlands and swamps. Grass-trees (*Xanthorrhoea australis*) send tall, slender spires aloft, each bearing thousands of creamy, nectar-filled flowers to entice the honeyeaters. The Peppermints (*Eucalyptus dives*) burst into blossom, filling the air with warm fragrance and bees. The heathlands of the Grampians are perhaps the richest in the state, and contain one of its rarest and most unusual plants, *Calectasia cyanea*, commonly called the Blue Tinsel Lily. Found also in sandheaths of South Australia and Western Australia, its metallic purple-blue stars gleam among the tangle of wiry stems, leaves, twigs and bush litter which carpets the ground.

Exceptionally beautiful are the many members of the heath family (Epacridaceae) found in the Grampians. Knee-high mounds of *Astroloma conostephioides*, the Flame Heath, burst into a scarlet sheen of flower, which lasts from spring to summer. Compact, sometimes prostrate, clumps of *Styphelia adscendens*, the Golden Heath, bloom continuously from winter to summer, their furry petals gently curled back around the long stamens. Commonly found on sandstone ledges is the endemic form of Common Heath (*Epacris impressa var. grandiflora*), a tall, robust form of Victoria's floral emblem, which is covered with rosy bells from winter to spring.

Another Grampians endemic is *Thryptomene calycina*, which colonises moist sandy depressions, mainly in the northern region. Early spring brings a glory of white and palest pink to these shrubs, growing thickly on stream sides, in the dappled forest shade.

Baeckeas, calytrix, grevilleas, bauera, boronias, micromyrtus, orchids, leptospermums, melaleucas, and acacias flourish in the sandy soils, a wealth of delicate beauty and colour extending from the valley floors to the wooded slopes, from the craggy bluffs on to the peaks and plateaus.

Grevillea alpina
Glorious in its combination of gold and orange flowers, this form of *Grevillea alpina* is found in the sandy, shaded woodlands of the Grampians in western Victoria. It grows to perhaps half a metre in height and about the same in width, often with its wiry stems entangled amongst heaths, correas and small pea plants. Rarely without some flowers, with a peak of blooming during winter and early spring, it is especially attractive to honeyeaters and has found ready acceptance in many gardens.

Holly grevillea *Grevillea aquifolium*
This is one of several grevilleas found in the Grampians, and is variable within the region. Endemic to western Victoria, it is also found in the Little Desert and south around the Lower Glenelg. The typical Grampians form grows as a shrub to about 2 metres, dense and compact, with stiff holly-shaped leaves of grey-green. Mainly spring-flowering, the bright scarlet and gold 'toothbrushes' attract bees and honeyeaters in profusion. Mostly it is found in deep sandy soils, often in association with woodland but also as an open-heath component, and ranging in habit from the tall shrubby form to a dense, entirely prostrate mat.

Rosy Bush-pea *Pultenaea subalpina*
The only member of this genus to bear flowers of this colour, the Rosy Bush-pea is confined to the Major Mitchell Plateau, part of the precipitous eastern flanks of the Grampians. Flowering in late spring to early summer, the shrubs, less than a metre high, form a dense thicket amongst the woodland in a few gullies which drop from the lip of the plateau. It is a very beautiful and rare sight, well worth the full day's walk needed to reach the only area from which it is known.

Blue Tinsel-lily *Calectasia cyanea*
The only member of its family, Calectasiaceae, the Blue Tinsel-lily (opposite) is quite un-lily-like. Spiky leaves are crowded along its wiry stems, which bear clusters of single, very spectacular flowers. Six deep gold anthers jut from the centre of six metallic, gleaming purple-blue segments, spread evenly around. A rare plant, known only from a few localities in western Victoria, south-eastern South Australia, and Western Australia, it inhabits low sand-heaths and, unless in flower, could easily be missed, for it blends itself thoroughly with the heathland plants.

The Stirling Range

The Stirling Range, in southern Western Australia, is a superbly wrought natural garden. Ancient peaks rear skyward, with Bluff Knoll over 1100 metres; bold slopes, thickly clothed with dense vegetation, hide the delicate tracery of wildflowers for which the region is renowned, and which closer investigation reveals in magical variety. Woodlands, heath and low shrubbery contain more than 500 species, about half of which belong to three families — Proteaceae, Myrtaceae and Fabaceae. Many species are endemic to the Stirling Range, and some confined to one locality only within the Range.

Set amidst vast stretches of farmland, the age-old Stirling Range thrusts its rugged bulk to the sky. As one approaches, each peak — tantalisingly mysterious from afar — becomes prominent and distinctive. The summit of the highest peak of all, Bluff Knoll, is often wreathed in a swirl of cloud. The slopes of these ranges are thickly clad with heathland species. Many have tough, rigid or leathery leaves, enabling them to withstand periods of dryness and summer heat, for although it is an area of high rainfall, there is little permanent water, and the creeks run only after rain.

Both rare and endemic species flourish within the Stirlings which, in typical sandstone fashion, support one of the richest shows of flora in Australia. The dainty darwinias, or Mountain Bells, may be the best-known plants of the region. *Darwinia macrostegia*, the Mondurup Bell, with waxy, painted red and white bells, is confined to Mondurup Peak. On Bluff Knoll, two species are readily found: the pale green *Darwinia collina*, and the delicate pink *D. squarrosa*. *D. leiostyla*, with deep pink nodding bells, is more widespread.

But the flora is exciting and captivating wherever one looks. Beaufortias, brush-flowered members of the family Myrtaceae, are confined to Western Australia, and many occur on the heathlands and rocky slopes of these ranges. Particularly striking is *B. decussata*, a tall, slender shrub with symmetrical foliage and large scarlet brushes, while the smaller *B. heterophylla* forms compact mounds, its arching branches bearing tiny, fluffy brushes of deep red.

Coneflowers and Mountain Paper-heath
Isopogon latifolius and *Sphenotoma drummondii*
Flowering together in late spring, these shaggy pink Coneflowers and the white-headed Mountain Paper-heath are found only in the rocky crevices of the highest peaks in the Stirlings. *Isopogon latifolius* is a sturdy shrub of the Proteaceae family, its broad leaves bearing a strong resemblance to many of the South African Proteaceae members. Each flower head is composed of many individual flowers, 'mop-like' in arrrangement and silken in texture. Tufted and spiky, bearing its clustered flowers at the end of each stem, the Paper-heath belongs to the family Epacridaceae, and is one of six species of *Sphenotoma*, all endemic to Western Australia. Its common name refers to the texture of the petals, which become papery with age.

Banksia coccinea
Scarlet and grey, the cones of this magnificent banksia are a wonderful sight during September and October. Whether a single specimen rising above the carpeting heathlands, or in orchard-like proportions on the south coast east of Albany, its beauty is unrivalled. Mostly it grows to 2 or 3 metres, with a number of slender branches rising from near its base. Compact grey leaves crowd along the stems, each terminating in the flattened flower spike.

A number of banksias are prominent in the Stirlings. *Banksia solandri* will be found on the slopes of the eastern peaks, its soft, furry bronze new growth and velvety brown flowers a delight. Growing in dense thickets on the heathlands, *B. coccinea* bears distinctive scarlet and grey cones on stout grey-leaved stems. Prostrate species such as *B. repens* and *B. prostrata* grow along the valley floors, their stems running along the sand, leaves erect and the new buds of cherry-pink and bronze nestling amongst the foliage.

Closely related, dryandras are confined to Western Australia and are common throughout the south-west. Attracting as much attention as the banksias, the range of foliage, flower and habit within this genus is astonishing. Aptly named Honeypots, many bear their nectar-rich, golden or bronze flowers deep within their protective leaves. Insects and honeyeaters, however, find the flowers easily, and in seeking the nectar, act as pollinators. *Dryandra formosa* will be readily recognised on the slopes of Bluff Knoll, its finely serrated leaves surrounding the deep gold flowers.

Frequenting the same precipitous faces, and flowering with *D. formosa*, is the magnificent *Isopogon latifolius*, its broad leaves and large deep pink cone-flowers a striking picture in the rocky clefts.

The drier, semi-shaded, open woodlands are the habitat of many delicate and beautiful wildflowers. The dainty purple bells of *Platytheca verticillata* may be found in moist depressions among the open forest of the valleys, together with numerous orchids, droseras and stylidiums.

In the heathlands, carpeting the outlying country and merging into the woodlands, one finds a wonderful array of species. Closely packed, their stems and leaves entwined and tangled, are hibbertias of gold and orange, isopogons of soft pink, conospermums of blue and grey, acacias and hakeas, dryandras, banksias and beaufortias, and pea flowers of several genera.

Rising above the low, dense heath, imparting a timeless quality to the landscape, are the black-trunked *Kingia australis*, or Black Gins, their silver skirts topped with the knobby drumsticks which are the fruiting cones.

Each of these landscapes, uplifted and weathered to craggy escarpments, sheer cliffs, broad shelves and sweeping slopes, supports a flora of extraordinary diversity and beauty.

Mountain Bells *Darwinia leiostyla*
The Mountain Bells of the Stirling Range (opposite) are a fascinating study. Many appear to have evolved on, and remained confined to, just one peak. *Darwinia macrostegia*, with painted red-and-white bells, is isolated on Mondurup Peak; the green-belled *Darwinia collina* and its pink-belled twin *Darwinia squarrosa* are found only on Bluff Knoll. *Darwinia leiostyla*, however, is a little more widespread through the region. It grows as a small rounded shrub to 60 cm, with attractive heath-like foliage. Exquisite rosy bells nod over the entire bush throughout spring. The outer 'petal-like' bell is actually composed of bracts, which enclose a tiny cluster of flowers. Darwinias belong to the family Myrtaceae, and are known from both western and eastern sides of the continent. Not all are bell-like; several bear their flowers in upturned bundles.

Showy Dryandra *Dryandra formosa*
Perched in rocky crevices and clinging to the precarious upper slopes of Bluff Knoll, as well as on coastal heaths and sandy hills to the south, *Dryandra formosa* (above) is one of the most spectacular of its genus. Soft, finely-serrated foliage surrounds the deep golden-bronze cones; many dryandra flowers have a gleam and lustre which is almost metallic. Members of the Proteaceae family, dryandras are confined to Western Australia, and almost all occur in the south west.

Black Gin *Kingia australis*
Rugged cliffs of the Stirling Range drop sheer to the sloping ridges, clad densely in a thick, almost impenetrable tangle of shrubbery and heath. A silver-haired *Kingia* stands motionless as if gazing into the ancient past, as the pale sun gleams through the mists. Kingias belong to the same family as xanthorrhoeas (Xanthorrhoeaceae), but instead of one long flowering spike, they produce several knob-ended 'drumsticks' which bear first the flowers, then the seeds.

Tribulus occidentalis
Commonly found in the north-west of Western Australia, and across the desert to South Australia and New South Wales, this species (left) grows on gravelly or sandy soils. Generally a prostrate herb, it may be either ephemeral or perennial, and belongs to a family known as Zygophyllaceae, which contains nine Australian species of *Tribulus*. The hairy trailing stems bear grey-green pinnate leaves, and the silky buttercup-yellow flowers may bloom at any month of the year, depending on rainfall. Round, hard, spiny seed capsules succeed the flowers.

Arid Australia

Sturt Desert Pea *Clianthus formosus*
Glory of the inland plains, the Sturt Pea creeps its soft stems over gravel and sand, spreading a mat of furry silver leaves among spinifex and tussock. The gleaming flowers, each centred with a glistening boss of jet-black or crimson, are bunched around an upright stem, like so many enquiring gnomes gazing about. Belonging to the pea family, Fabaceae, *Clianthus formosus* is classed as an ephemeral herb; good rains early in the year mean a massed display from June to September. The Sturt Pea, floral emblem of South Australia, was named after the explorer Charles Sturt. It was one of the first Australian wildflowers collected by European man, recorded by William Dampier in 1699.

The land form of Australia's interior captures the senses in timeless suspension. Nothing disturbs the serenity of this ancient landscape, weathered over millions of years to bony, ribbed ranges, rolling plains and distant horizons. At night, the velvet blue-black canopy is radiant with frost-glittering stars, and the whispering sigh of the mulgas ruffled by a gentle breeze stirs the silence.

Winter and summer are the only seasons: the spring and autumn of softer climates have no place here. Winter may be bitter-cold, the mornings heavy with frost, but the day which follows such a dawn is clear and sunny and of gentle warmth. Summer is unquestionably hot.

Occasionally, there may be storms — dust-storms, perhaps, which bring a rain of falling red sand; or rainstorms, heralded by billowing black clouds and wierd light, bursting upon the thirsty earth in deafening torrents.

Then magic follows. Carpets of pink, white, mauve and yellow flood over the soil, a sea of waving, quivering colour which, on close inspection, reveals uncountable flowers. Ephemerals, they are called. Many belong to the daisy, or Asteraceae family. Their tiny seeds have lain buried in the soil, perhaps for years, patiently awaiting their chance to sprout in damp warmth, to flower briefly, to set and shed their seed, and to wait again for those conditions which trigger the cycle.

Little red will be seen but for the bold scarlet of *Clianthus formosus*, the Sturt Desert Pea. Its soft grey pinnate leaves and furry stems creep over the gravel and sand, supporting clusters of glistening flowers, some showing a jet-black central boss, others a deep cherry-red.

Nowhere in the entire arid zone is there a desert of the truly barren type. There is, in fact, a great deal of vegetation to be seen, even in the Simpson Desert, which receives the least rainfall of all.

Roughly, the vegetation can be categorised into four broad groups: woodlands, savanna grasslands, spinifex country, and succulent (or saltbush) steppes.

The woodlands range from the stands of great spreading River Red Gums (*Eucalyptus camaldulensis*) which line the watercourses, to mulga country and mallee. Growing with their roots deeply buried in the sandy edges of the rivers, which are often dry, the Red Gums live to a great age. Able to withstand severe droughts, by sending their roots down far enough to tap underground moisture, they require occasional flooding to stimulate regeneration. Thick little forests of seedlings sprout following a prolonged deluge, the stronger ones surviving as the new generation. The Coolibah, *Eucalyptus microtheca*, spreads out over the floodplains, similar to, but not as large as the Red Gum.

A particular characteristic of inland trees is their light-coloured trunks. Many, in fact, are a startlingly pure white, with a white powder which will rub off on the hand, and which

Velleia rosea
One of Western Australia's many dainty herbs which spring up after rain, *Velleia rosea* (above) carpets the red loam beneath open mulga woodland. The delicate pale pink five-petalled fan flowers appear during spring, borne on slender stems rising from a rosette of leaves, and imparting a transient softness to the land. Belonging to the family Goodeniaceae, eleven species of *Velleia* are known from Western Australia, with a total of twenty species found in Australia and New Guinea. Easy to wander through (except for avoiding trampling the flowers), the mulga woodlands display a gentleness which vanishes in the heat of summer winds and sun.

Eucalyptus brevifolia
Smooth and powdery white, the rounded trunk of *Eucalyptus brevifolia* (opposite) rises from a golden sea of everlastings. After good rains, the red loam of the inland disappears beneath a mantle of colour. Ephemeral herbs, mainly of the daisy family, germinate, flower and set seed almost in a matter of moments, ensuring their cycle of regeneration. The intense light of northern Western Australia is reflected in everything — sky, earth, flower and tree — flung back to dazzle the eye and fill the air with brilliance.

is associated with the tree's ability to withstand the heat and glaring light. The Ghost Gum of the Centre, *Eucalyptus papuana*, and the smaller *Eucalyptus brevifolia*, stand in dazzling contrast against the vivid orange-red rock and sand.

Mulga, *Acacia aneura*, mostly grows in pure stands covering vast areas of flat or gently undulating country. Often the sand supports a mass of small ephemerals and low shrubs. Mulga country is easy to walk through, and is good for camping. Not only Mulga, but numerous species of acacias will be found in the dry inland of Australia, superbly adapted to the dessicating conditions. Casuarinas, too, are part of the typical woodland landscape, their fine needle foliage exposing the least possible surface area to the atmosphere, and their trunks clothed in thick bark.

The mallee areas are widespread, and contain many species of specially adapted eucalypts, characterised by a large woody base or ligno-tuber, from which sprout numerous branches. The flowers of these species are often spectacular, either in massed groups, or individually large. The gumnuts which follow the flowers are equally appealing in their shape and texture. Growing in association with the eucalypts is a wide range of smaller shrubs, particularly melaleuca and acacia species, and many brightly-flowered herbs and ground-covering plants.

Grasslands, both savanna and spinifex, cover enormous tracts of country. Savanna plains mainly support perennial grasses such as Mitchell grass (*Astrebla* species), and spinifex is a term applied to several species of grass, including *Triodia* and *Plectrachne*. Typically, they form dense, extremely prickly mounds of considerable size, and tend to die out in the centre, thus forming rings of growth. While the savanna plains tend to be treeless, spinifex grasslands are generally found in association with either mallee, mulga or scattered shrubs.

The succulent steppes might, to the casual eye, appear to be all saltbush. In fact, three particular groups of species would be apparent on closer investigation: saltbush, bluebush and samphire, each associated with different soil types, and all belonging to the family Chenopodiaceae.

The saltbush group, mainly of the genus *Atriplex*, grows on soil which is often weakly saline; while bluebush communities, of predominantly *Maireana* species, will be found on limestone, mostly growing as low and spreading shrubs. The soft greys, greens and blues of their foliage blend into a muted sea, either as an undershrub in woodland or, as on the Nullarbor Plain, spreading treeless to the horizon.

Samphire communities, of various genera but particularly *Halosarcia*, are characterised by fleshy, jointed stems, and are found on highly saline soils in depressions around or in salt lakes. Such communities rarely contain shrubs or trees.

A number of features characterise the vegetation of the arid region. In particular, the sclerophyll (or hard-leaved) nature of most of the plants is one of the most effective means of reducing moisture loss through foliage. In some cases, the leaves are almost non-existent, as in the casuarinas, or they are narrow and leathery, presenting as little surface area as possible to the sun.

Rosy Sunray *Helipterum roseum*
A dainty annual herb, the Rosy Sunray has, like all helipterums, delicate papery flowers, often known as 'everlastings'. Deep rose-pink, palest shell-pink and white, the flowers open their faces to the spring sunshine, closing in the evenings or on grey cloudy days. Growing only 30 cm high, on fine green stems, they carpet the sandy plains of the inland regions of south-west Western Australia, flowering prolifically and shedding their soft seed in rapid succession to complete their life-cycle.

The foliage may be greatly reduced in size or quantity, or it may be fleshy, or covered with a film of closely matting hairs which insulate the leaf. Sticky or warty surfaces, and the powdery bloom or 'glaucous' nature of the leaf and bark of certain eucalypts especially, are other adaptions which protect the leaf and enable survival.

Succulence is not a striking feature of Australia's inland plants. While it is highly developed in most of the world's desert flora, notably in the cactus family, only in *Chenopod* communities does it dominate, together with some relatively minor groups such as *Calandrinia* and *Portulaca* species.

Australia's arid zone, covering about one third of the total continent, is a land shaped by time. Elemental, and uncompromising in its nature, its beauty is not for the fainthearted. Only the true survivor can stay alive — whether plant, animal or human. Evolution has furnished this environment with perfectly adapted species, and it is interference by the ignorant which causes imbalance and its inevitable price — waste, desolation, death.

Parakeelya *Calandrinia remota*

Calandrinias, or Parakeelya as they are commonly known, are small, fleshy, prostrate or low-growing herbs of the family Portulacaceae. Widespread in the arid regions of Australia, they splash the hot sand with vivid pinks and purples. Each flower lasts for just a day, but is succeeded by another so that flowering continues over several months of late winter and early springtime. *Calandrinia remota* is one of eighteen species recorded for central Australia; rising from a basal rosette of succulent leaves, slender stems bear five-petalled glossy flowers many centimetres above the sand. Only a few groups of plants in Australia exhibit the characteristic of succulence. Most species have adapted to the hot dry conditions of the arid inland by developing a sclerophyll or hard-leaved nature.

Simpson Desert
Nowhere in Australia's arid region is there a desert entirely without vegetation — not even in the Simpson Desert, which receives the lowest rainfall of all the continent.

Golden Long-heads *Podotheca gnaphalioides*
The simplicity of the Australian landscape lies mutely expressed in this vast tract of flower, a golden plain stretching to an endless horizon. *Podotheca gnaphalioides* is found over much of Western Australia, and is one of the arid zone's many ephemerals (short-lived plants) in the daisy or Asteraceae family. Individually, each tiny flower head is composed of even tinier florets. En masse, millions of flowers spread a wave of colour across the earth, lasting for a few brief weeks.

Glossary

ANTHER, pollen-bearing part of a stamen.

BRACT, a modified or reduced leaf.

ENDEMIC, confined to a specific geographic region.

FLORET, a very small flower.

FOLLICLE, a fruiting capsule, splitting laterally.

LABELLUM, a lip; refers to a distinctive petal as in orchids.

LANCEOLATE, shaped like a spear-head.

PEDICEL, the stalk of a single flower.

PERIANTH, a combination of calyx (sepals combined) and corolla (petals as a whole).

PETIOLE, the stem of a leaf.

PHYLLODE, a modified, flattened leaf stalk, especially of acacias.

PINNATE, describes a leaf composed of pairs of leaflets arranged along a stalk in a feather-like manner.

RACEME, a flower-head bearing a number of flowers on pedicels of equal length.

SEPAL, part of a calyx, usually green and leaf-like.

SCEROPHYLLOUS, hard-leaved.

STYLE, tubular floral part terminating in the stigma which receives the pollen.

TEPAL, as in grevilleas where the main floral segments are fused into one unit.

Index

Acacia
  cultriformis 6
  denticulosa 7
  drummondii 7
  pravissima 6
  retinodes 4
  terminalis 5
Actinotus helianthii 49
Albany Bottlebrush 34
Anigozanthos manglesii 46

Banksia
  baxteri 8
  blechnifolia 9
  coccinea 59
  hookerana 9
  praemorsa 11
  prostrata 11
  serrata 10
  spinulosa 10
Bell-fruited Mallee 17
Blandfordia grandiflora 51
Blandfordia nobilis 51
Blue Tinsel-lily 57
Bogong Daisy-bush 29
Boronia serrulata 49

Calandrinia remota 67
Calectasia cyanea 57
Callistemon speciosus 34
Calocephalus brownii 37
Calochilus robertsonii 39
Caltha introloba 27
Carpobrotus rossii 37
Celmisia asteliifolia 26
Chiloglottis gunnii 38
Christmas Bell 51
Clianthus formosus 63
Coneflower 59

Darwinia leiostyla 61
Deciduous Beech 1
Dryandra formosa 60

Epacris impressa 42
Epacris longiflora 50
Eriostemon australasius 52
Eriostemon verrucosus 40
Eucalyptus
  brevifolia 65
  caesia 14
  leucoxylon 13
  macranda 15
  macrocarpa 16
  preissiana 17
  regnans 12

Fairy Waxflower 40
Flannel Flower 49
Fuchsia Heath 50

Golden Long-heads 69
Grass-tree 54
Grevillea
  alpina 55
  aquifolium 55
  bipinnatifida 19
  buxifolia 19
  juniperina 20
  robusta 20
  sericea 18
  victoriae 21
Groundsel, Variable 36

Hakea multilineata 3
Hardenbergia violacea 41
Helichrysum acuminatum 28
Helipterum roseum 66
Hibbertia procumbens 44
Holly Grevillea 55
Hovea purpurea 31

Isopogon latifolius 59

Kangaroo Paw 46
Kingia australis 60
Kunzea ericifolia 24

Lechenaultia formosa 35

Mangrove 32
Marsh Marigold 27
Mountain Ash 12
Mountain Bell 61
Mountain Paper-heath 59

Native Violet 42
Neopaxia australasica 30
Nothofagus gunnii 1

Olearia frostii 29
Orchid, Bird 38

Parakeelya 67
Persoonia pinifolia 53
Podotheca gnaphalioides 69
Pultenaea paleacea 44
Pultenaea pedunculata 45
Pultenaea subalpina 56

Ranunculus anemoneus 25
Ranunculus graniticola 31
Rhizophora stylosa 32
Rose of the West 16
Rosy Sunray 66

Scleranthus biflorus 31
Senecio lautus 36
Silky Oak 20
Snow Daisy 26
Sphenotoma drummondii 59

Stackhousia pulvinaris 25
Sturt Desert Pea 63
Styphelia strigosa 42
Sydney Rock Rose 49

Telopea speciosissima title page
Tetratheca ciliata 42
Thysanotus tuberosus 41
Tribulus occidentalis 62
Trigger Plant 23

Velleia rosea 64
Viola hederacea 42

Waratah title page

Xanthorrhoea australis 54